A HANDBOOK OF ORGANIC CHEMISTRY PRACTICAL

(*TESTS, PROCEDURES AND RESULTS*)

FOR TERTIARY INSTITUTIONS

INNOCENT C. ONUNKWO (M.Sc., PDE, B.Sc.)

i

Publisher:
XS Publishers

DEDICATION

The book is solely dedicated to God,

who made it a reality.

TABLE OF CONTENTS

Identification, extraction and purification of organic compound:

Introduction

1. Preliminary Identification Tests

    i. Colour, odour and physical state
    ii. Solubility
    iii. Unsaturation
    iv. Aromaticity

2. Extraction
    i. Simple distillation

3. Chromatography
    i. Paper Chromatography

4. Phytochemical analysis: preparation and test
    i. Test for carbohydrate and glycosides (Molish test)
    ii. Test for Tannin
    iii. Test for Saponin (Froth test)
    iv. Test for flavonoid
    v. Test for alkaloids
    vi. Test for sterols and terpenes (Libermann's and Salkowski tests)
    vii. Test for anthraquinones and glycosides

(d) Determination of percentage yield

(e) Preparation of solution by dilution

(f) Preparation of molar solution of substance (liquid)

INTRODUCTION

Organic Chemistry is the study of hydrocarbon or carbon containing compounds in terms of its structure, properties, reactions and preparation. This study is not limited to carbon containing compounds as some other elements such as hydrogen, nitrogen, sulphur, oxygen, silicon and phosphorus are also studied in relations to carbon. Organic chemistry is broadened from natural to synthetic or man-made organic compounds, which has found its use in production of pharmaceutical, paint, cosmetic, food, plastic, petroleum and explosive products in many manufacturing laboratories and industries.

The practical aspect of organic chemistry is the ideal for studying organic chemistry as it gives an insight of the skeletal and cognitive understanding of the principles of organic chemistry. These principles or skills are needed by science students and research scientists for better understanding of scientific works.

It is of great importance that this academic book should be circulated across different Libraries, especially in tertiary institutions and be recommended to science students and researchers in order to aid and equip them with necessary scientific, laboratory and practical skills needed to carry out their scientific researches.

CHAPTER

ONE

IDENTIFICATION, EXTRACTION AND PURIFICATION OF ORGANIC COMPOUND: INTRODUCTION

Identification, extraction and purification of organic compounds can be carried out by preliminary test analysis, extraction, chromatography and phytochemical.

1. **Preliminary Identification Tests**

    (a) Colour, physical state and odour

    (i) **Colour and physical state** – have important role to play in preliminary identification of organic compounds. For examples, naphthols are usually pink in colour and most of them are solid in nature; nitro compounds (e.g. nitrobenzene) are yellow in colour. Some colourless liquid organic compounds include alcohols, aldehydes, ketones, simple esters, lower ethers and aromatic hydrocarbons. The colour and physical state of some organic compounds may change on long storage due to climatic and environmental factors.

    (ii) **Odour** – the odours of organic compounds also help in their identifications. For example, amines possess fish-like smell; nitrobenzenes possess almond-like smell; acetamides and acetonitriles possess rat-like smell, etc.

    The odour of organic compounds may change with time due to storage and environmental factors.

2

(b) Solubility

Solubility of organic compounds also helps in identifying if the compound is acid or basic or neutral. Some of the solutions used to check the solubility are water (H_2O), dilute Hydrochloric acid (HCl), dilute Sodium hydroxide (NaOH) and dilute Sodium carbonate (Na_2CO_3). Water is used to check the solubility by pH analysis of the dissolved compounds in order to determine the pH scale of the solution; whether it is acidity (1 - 6.9), basicity (7.1 -14) or neutral (7.0). Acidic compounds are insoluble in HCl, soluble in NaOH and produces effervescence with Na_2CO_3 while basic compounds are soluble in HCl, insoluble in NaOH and do not produce effervescence with Na_2CO_3. Neutral compound are either soluble or insoluble in HCl and NaOH.

(c) Test for Unsaturation: it can be done using -

(i) Baeyer's test

Procedure:

- Make a 5 mL each of the solutions of saturated and unsaturated compound in test tubes A and B respectively.
- Add few drops of alkaline Potassium permanganate ($KMnO_4$) in each of the solutions, shake well and observe.

Result:

- Solution A shows no decolourization with $KMnO_4$ while solution B (alkene and alkyne) undergoes addition reaction with $KMnO_4$ to produce colourless diols, that is aliphatic unsaturation changes the purple colour of $KMnO_4$ changes to brown MnO_2.

$$CH_2{=}CH_2 + KMnO_4 \longrightarrow CH_2\bar{O}{-}CH_2\bar{O} \longrightarrow CH_2OH{-}CH_2OH + MnO_2$$

Aromatic unsaturated compounds do not react at all or slowly with $KMnO_4$ and show no decolourization.

(ii) Bromine water test

Procedure:

- Make a 5 mL each of the solutions of saturated and unsaturated compound in test tubes A and B respectively.

- Add few drops of Bromine water in each of the solutions, shake well and observe.

Result:

Solution A shows no decolourization with Bromine water while solution B (alkene and alkyne) undergoes addition reaction with Bromine water to produce colourless halohydrine; that is, aliphatic unsaturation reacts with Bromine water to produce halohydrine. $CH_2{=}CH_2 + HOBr \longrightarrow CH_2OH{-}CH_2Br$

Unsaturated aromatic amines and phenols produce insoluble white precipitates of cyclic bromonium ions with bromine water and shows decolorization due to aliphatic unsaturation.

(iii) Addition of Bromine Chloroform ($CHCl_3$) or Carbon Tetrachloride (CCl_4)

Procedure:

- Make a 5 mL each of the solutions of saturated and unsaturated compound in test tubes A and B respectively.
- Add 2 mL of $CHCl_3$ or CCl_4 in each of the solutions, shake well and observe.
- Add few drops of Bromine water in each of the solutions, shake well and observe.

Result:

- Decolourization of solution B to produce colourless dibromo compound without evolution of Hydrogen Bromide (HBr) while solution A shows the release of HBr.

$$CH_2{=}CH_2 + \text{Bromine water} \xrightarrow{CHCl_3/CCl_4} CH_2Br + CH_2Br$$

- Rapid declourization shows presence of aliphatic unsaturation.

(iv)   Test for Aromaticity

Procedure:

- Put small amount of organic compound in a streak and burn in a Bursen burner flame.

Result:

- If the compound is aromatic, it will burn with a sooty flame (black smoke), showing high percentage of carbon present. That is, incomplete oxidation of carbon. Long chain aliphatic contain lots of carbon compounds and therefore also burn with a smoky flame.

2. Extraction

Extraction is a method of separation that utilizes the differences in solubility of the materials present. Extraction process depends on chemical and physical characteristics of the targeted compound, availability of the plant sample, eco-friendly approach and yield of the extract. There are different extraction methods which includes:

(a) **Maceration**: It is a simple, less efficient and time consuming method used for extraction of thermolabile components; example, teabag in water.

(b) **Percolation**: It is more efficient than maceration process and results from the passage of liquid droplets through solid material conducted at room temperature.

(c) **Decoction**: Result in heating and boiling plant extract in water. It is not a good process for volatile or thermolabile components.

(d) **Reflux extraction**: Used for extraction of volatile components but not for thermolabile natural products and it is more efficient than percolation and maceration, and requires less time and solvent.

(e) **Soxhlet extraction**: is an automatic and continuous extraction method which utilizes the principle of reflux and siphoning to continuously extract sample (macerated) with fresh solvent percolation. It is an efficient method of extraction and requires a lot of solvents.

(f) **Pressure Liquid Extraction**: Uses high pressure to keep solvent liquid above boiling point for extraction of lipid solutes – which has low solubility and diffusion. It has efficient solvent and less time utilization.

(g) **Pulse Electric Field Extraction**: It destroys membrane structures to increase mass transfer using non thermal, pulse field strength and specific energy for extraction.

(h) **Supercritical Fluid Extraction**: Used to extract thermolabile components using supercritical fluid (e.g S-Carbondioxide) with good solubility and diffusivity to liquid and gas respectively.

(i) **Ultra-Assisted Extraction**: Promotes the dissolution and diffusion of solutes through acceleration and heat transfer to achieve efficient extraction. It is used for extraction of the thermolabile and unstable compounds using ultra-sound wave energy.

(j) **Enzyme Assisted Extraction**: Achieved through the use of hydrolytic action of enzymes on the component of the membrane and macromolecules inside the cell to facilitate the release of the natural product during extraction.

(k) **Hydro distillation and steam distillation**: for extraction volatile oil substances; example, essential oil.

(i)   Simple Distillation method

This is used for separation of liquids (miscible) based on the differences in their boiling points whereby the substance with lower boiling point evolves as the boiling temperature is reached before the substance with higher boiling point. This method can also be used for reaction of substances; example, Reduction of Cyclohaxanol (162°C) with hydrogen phosphate (H_3PO_4) under heat to produce Cyclohexene (83°C).

An illustration of Simple distillation method for example is the separation of the mixture of ethanol (boiling point at 78°C) and water (boiling point at 100°C).

Materials required:

Distillation flask, hotplate, thermometer, hosts for running water, condenser, distillate flask, cork, and clamp.

Reagent: water and ethanol.

Procedure:

Put equal mixture of ethanol and water into the distillation flask, connect the thermometer and the mouth of the condenser the cork and cork the distillation flask and make sure it is air tight.

Insert distillate flask at the other lower end of the condenser for collection of the distillate. Connect the hosts to the condenser sides where there are openings, one the end where water will flow into the condenser from the tap and the other where water will flow out of the condenser. Clamp all tightly. Switch the tap ON and heat the distillation flask slightly above 78°C but not close to 100°C while regulating the temperature with the aid of the inserted thermometer until all the ethanol have evolved and collected.

Result:

The pure ethanol is collected with the distillate flask while water remains at the distillation flask.

Precaution:

- Distillation flask should not be more than half full.
- Thermometer should be correctly placed to measure the vapour temperature not the liquid mixture.
- Heat source should be easily removable to regulate the temperature.
- The glassware should be tightly clamped.
- Distillate collection should be done at the right boiling point to avoid contamination.

(ii)   Fractional Distillation: This is a more sophisticated method of distillation that is used when there are mixtures of more than two different substances whereby the substances are heated at high temperature and each separates in a fractional column based on the differences in their boiling points. An example of this, is the fractional distillation of crude oil.

3. Chromatography

This is a biophysical technique which enables the separation, identification and purification of compounds present in mixture of a qualitative and quantitative analyses. The compound to be separated are distributed into two phases: the stationary and mobiles phases. The two types of chromatography are – The Column chromatography which is used in High Performance Liquid Chromatography (HPLC), Gas Chromatography (GC), etc., and Planar Chromatography – used in Paper Chromatography, Tin-Layer Chromatography, etc.

(i) **Experiment on Paper Chromatography** – it is a Liquid-Liquid Chromatography sub-divided into ascending, descending, circular and 2-dimensional types of paper chromatography. This experiment will cover only on ascending paper chromatography.

Materials required:

Filter paper (cellulose), solvent (Petroleum ether mixed with Acetone in 9:1), mortar, pestle, scissors, chromatography jar with lid, pencil, thread, capillary tube, glass funnel, conical flask, measuring cylinder, ruler and fresh green leaf.

Procedure:

- Take small amount of clean fresh green leaf into a mortar.
- Put small amount of equal Petroleum ether and Acetone mixture and grind well.
- Filter the extract and collect the supernatant using glass funnel fitted with cotton wool or folded filter paper in a conical flask.
- Cut a rectangular filter paper with scissors and use the pencil and ruler to draw 1 cm^3 line from the shorter end.
- Use the pencil to spot the middle from the drawn line and use the scissors to make a V-shape from the edges of the drawn line.

- Use the capillary tube dipped in the extract to spot the center of the pencil line severally and allow to air dry.
- Make a fold on the other end of the paper and place on the chromatography jar in the middle, held up by a cross thread.
- Remove the paper and pour the mixture of the solvent in such a way that it can only touch the pointed end of the paper and not the spotted extract (up to about 2 cm^3 deep).
- Carefully place the spotted paper into the jar with a lid.
- Allow to stand until the solvent reaches above the separated pigments and remove the paper and mark the end of the solvent (solvent front) travel, and spot the positions of each of the different colour of the separated pigments on the paper.
- Identify the pigments present on the green leaf extract based on the separated colour.

Result:

- Green colour represents presence of Chlorophyll a.
- Yellowish green colour represents presence of chlorophyll b.
- Light yellow colour represents presence of xanthophyll.
- Deep yellow colour represents presence of carotene.
- Red colour represents presence of anthocyanin.
- Using meter rule to measure the distances traveled by each pigments and calculate the retardation factor (Rf-value). Rf =Distance traveled by pigment divided by distance traveled by solvent.

Precaution:

- Make sure the glassware used is washed thoroughly to avoid contamination of the sample.
- Make sure the leaf is as well washed with distilled water to remove dirt.
- Handle the solvent with care and avoid inhalation.

4. Phytochemical Analysis

This is the examination of plant sample to determine the chemical constituents that does determine the pharmacological activities and application of such plant sample. Some of the phytochemical constituents present in plant sample include Tannin, saponin, flavonoid, alkaloid, sterol, glycoside and anthraquinone.

Experiment on phytochemical analysis starts with plant sample preparation:
- Get a room dried and ground sample of plant.
- Divide the coarse powdered portion of the plant into large, medium and small size portion.
- Put the medium size in the beaker and small size portion in the test tube in a conical flask.

Aqueous Extract Preparation:
- Put 70 mL of distilled water to the medium size portion in the beaker and mix thoroughly.
- Put the aqueous extract mixture in the hot plate or heater until it boils.

- Filter off the boiled mixture in an empty flask fitted with glass funnel covered with cotton wool.

Acid Extract Preparation:

- Add 2 mL of dilute HCl to a test tube containing small coarse portion of the dry plant sample.
- Shake well, cover and leave for 20 minutes.
- After 20 minutes, filter in a cotton fitted filtration funnel into the beaker.

Alcoholic Extract Preparation:

- Add 80 mL of methanol in a beaker containing large portion of the coarse plant sample.
- Shake well, cover and allow to stand for 30 minutes.
- Use a cotton fitted glass funnel to filter into a beaker or porcelain dish.
- Slightly heat the beaker and allow to cool for complete evaporation of the solvent.
- Add 6 mL of Chloroform and scratch to get out the attached concentrate.

(i)   Test for Carbohydrates (Molish Test)

Procedure:

- Take 2 mL of the aqueous extract and put in a clean test tube.
- Add few drops of Molish reagent.
- Add slowly concentrated Tetraoxosulphate (IV) acid (H_2SO_4) in drops and excess.

Result:

- Fine ring appears on the bottom of the test tube for positive test.

(ii)   Test for Tannins

Procedure:

- Take 2 mL of the aqueous extract into a clean test tube.
- Add few drops of Iron (III) Chloride ($FeCl_3$).

Result:

- Blue colour shows hydrolysable tannins while green colour shows condensed tannins for positive test.

(iii)   Test for Saponins (Froth Test)

Procedure:

- Take 2 mL of the aqueous extract into a clean test tube.
- Cover and shake vigorously.

Result:

- Show up to 10 cm^3 froth which is persistent for 5 minutes for a positive test.

(iv)   Test for Flavonoids (using Acid Extract)

Procedure:

- Take 2 mL each of the acid extract and add separately to clean test tubes.
- Add 1 mL of distilled water to the first test tube.
- To the second test tube add 1 mL of Sodium hydroxide (NaOH).

Result:

- The first portion shows yellowish colour while the second test tube shows a dark colour for a positive test.

(v)   Test for Alkaloids (Dragendorff Test - using Alcoholic Extract)

Procedure:

- Use a capillary tube to take few drops of the methanolic extract and spot it on the filter paper.
- Spot another extract of a known alkaloid plant (for example, atropine) on the same filter paper some distance away.
- Spray the filter paper with Dragendorff's reagent.

Result:

- There would be pink colouration on the spots of the extracts for a positive test.

(vi)   Test for Sterols and Triterpenes (Libermann's Test)

Procedure:

- Take 2 mL of the alcoholic (methanolic) extract in a clean test tube.
- Add few drops of acetic anhydride.
- Add few drops of concentrated H_2SO_4 slowly from the wall of the test tubes.

Result:

- Reddish brown ring and green colouration at the upper layer of the test tube.

(a)   Test for Sterols and Triterpenes (Salkowski's Test)

Procedure:

- Take 2 mL of the alcoholic extract into a clean test tube.
- Add few drops of the concentrated H_2SO_4 slowly.

Result:

- A reddish brown ring appears on the lower layer of the test tube with no green colouration in the upper layer for a positive test.

(vii)   Test for Anthraquinones and Glycosides

Procedure:

- Take 2 mL of the alcoholic extract into a clean test tube.

- Add 1 mL of Aqueous Ammonia (NH$_4$OH) and shake.

Result:

- Rose-red colour in the upper layer of the test tube for a positive test.

CHAPTER

TWO

LASSAIGNE'S EXTRACT PREPARATION AND DETECTION OF ELEMENTS.

Lassaigne's extract is prepared and used for detection of elements present in organic compounds. The **preparation** is as follows:

Materials Required:

Organic compound, funnel, tripod stand, distilled water, wire gauze, Bunsen burner, fusion tube, test tubes, sodium metal, spatula, forceps, filter papers, china dish, glass rod and test tube holder.

Procedure:

- Take few grams of sodium metal with forceps and dry with filter paper.
- Put the dried sodium metal in a fusion tube in the flame of Bunsen burner to heat for few minutes until sodium metal melts.
- Take some quantities of organic compound using spatula into the fusion tube.
- Place the fusion tube over the flame of the Bunsen burner with the aid of test tube holder.
- Place the red hot tube into a China dish containing distilled water (Note, the tube will break).
- Crush the broken tube in the solution with glass rod.
- Heat the content of the China dish to boil and allow until the solution is reduce to half.
- Filter the content of the China dish in a clean test tube using glass funnel fitted with filter paper and clamped with tripod stand.

Result:

- The filtrate of Lassaigne's extract is obtained.

1. Detection of Nitrogen

Materials Required:

Ferric Sulphate fresh solution ($FeSO_4$), Ferric Chloride solution ($FeCl_3$), concentrated HCl, test tubes, test tube holder, Lassaigne's extract and droppers.

Procedure:

- Take 1 mL of Lassaigne's extract into a clean test tube using a dropper.
- Add 1 mL of $FeSO_4$ solution into the test tube containing the Lassaigne's extract with another dropper.
- Place the test tube over the flame of the Bunsen burner with the aid of test tube holder for few seconds.
- Replace to the test tube rack and add 1 mL of $FeCl_3$ (Ferric Chloride).
- Add about 1 mL of Conc. HCl into the test tube using another dropper.

Result:

- Sodium cyanide formed during the preparation of Lassaigne's extract is converted to Sodium ferrocyanide on treating with fresh solution of $FeSO_4$.
- On treating the solution of sodium ferrocyanide - $Na_4[Fe(CN)_6]$ with $FeCl_3$ and Conc. HCl, a blue coloured complex of sodium (ferric ferrocyanide), $Fe_4[Fe(CN)_6]^3$ is formed.

Precaution:

- Make sure the glass ware used is clean and different droppers are used for each of the solution.

- Handle the reagents with care to avoid hazards and inhalation.
- The Bunsen burner flame should be handled with care to avoid fire hazard.

2. Detection of Sulphur

(a)   Sodium nitroprusside test

Materials Required:

Sodium Nitroprusside solution, Lassaigne's extract, dropper, test tubes.

Procedure:

- Take 1 mL of Lassaigne's extract into a clean test tube using a dropper.
- Add 1 mL of sodium nitroprusside solution using another dropper.

Result:

- The sodium sulphide formed during the preparation of Lassaigne's extract gives a purple violet colour with $Na_2[Fe(CN)_5NO]$ (sodium nitroprusside) due to the formation $Na_2[Fe(CN)_5NOS]$ (sodium thionitroprusside).

(b) Lead acetate test

Materials Required:

Acetic acid, Lead acetate solution, Lassaigne's extract, test tubes and droppers.

Procedure:

- Take 1 mL of Lassaigne's extract into a clean test tube using dropper.

- Add 1 mL of acetic acid into the Lassaigne's extract solution using another dropper.
- Add few drops of Lead acetate solution using another dropper.

Result:

- Reaction of Lead acetate with sodium sulphide formed in Lassaigne's extract to produce a black precipitate of Lead sulphide.

3. Detection of Halogen (Cl, Br, I)

(a)   Silver nitrate test

(ai) Silver nitrate test for chlorine

Materials Required:

Concentrated Nitric acid (HNO_3), Silver Nitrate ($AgNO_3$), Ammonium hydroxide (NH_4OH), Lassaigne's extract, test tube, droppers, test tube holder and Bunsen burner.

Procedure:

- Take 1 mL of Lassaigne's extract into a clean test tube using dropper.
- Add slowly Concentrated HNO_3 with another dropper.
- Holding the test tube with test tube holder, place over the flame of the Bunsen burner for few minutes.
- Cool the test tube and add few drops of $AgNO_3$ solution.
- Add excess of NH_4OH solution and shake well.

Result:

- If the organic compound in the Lassaigne's extract contains sodium chloride ($NaCl$), then there will formation of white precipitate of silver chloride ($AgCl$) which is as a result of reaction of $NaCl$ in the Lassaigne's extract and $AgNO_3$ added. This shows the presence of Chlorine.
- The $AgCl$ precipitate dissolves in excess of NH_4OH solution to form soluble silver complex.

(aii) Silver nitrate test for bromine

Materials Required:

Concentrated Nitric acid (HNO_3), Silver Nitrate ($AgNO_3$), Ammonium hydroxide (NH_4OH), Lassaigne's extract, test tube, droppers, test tube holder and Bunsen burner.

Procedure:

- Take 1 mL of Lassaigne's extract into a clean test tube using dropper.
- Add slowly Concentrated HNO_3 with another dropper.
- Holding the test tube with test tube holder, place over the flame of the Bunsen burner for few minutes.
- Cool the test tube and add few drops of $AgNO_3$ solution.
- Add excess of NH_4OH and shake well.

Result:

- If the organic compound in the Lassaigne's extract contains bromine, sodium bromide ($NaBr$) present in the Lassaigne's extract would reacts with $AgNO_3$ to form of yellow precipitate of silver bromide ($AgBr$).
- The Silver bromide ($AgBr$) precipitate is sparingly soluble in excess of NH_4OH solution.

(aiii) Silver nitrate test for iodine

Materials Required:

Concentrated Nitric acid (HNO_3), Silver nitrate ($AgNO_3$), Ammonium hydroxide (NH_4OH), Lassaigne's extract, test tube, droppers, test tube holder and Bunsen burner.

Procedure:

- Take 1 mL of Lassaigne's extract into a clean test tube using dropper.
- Add slowly Concentrated HNO_3 with another dropper.
- Holding the test tube with test tube holder, place over the flame of the Bunsen burner for few minutes.
- Cool the test tube and add few drops of $AgNO_3$ solution.
- Add excess of NH_4OH and shake well.

Result:

- If the organic compound in the Lassaigne's extract contains iodine, sodium iodine (NaI) present in the Lassaigne's extract would reacts with $AgNO_3$ to form of yellow precipitate of silver iodide (AgI).
- The yellow precipitate of AgI is insoluble in NH_4OH solution.

(b)   Carbon disulphide test

(bi) Carbon disulphide test for bromine:

Materials Required:

Chlorine water, carbon disulphide (CS_2), dilute HCl, Lassaigne's extract, test tubes and droppers.

Procedure:

- Take 1 mL of Lassaigne's extract into a clean test tube using a dropper.
- Add 1 mL of CS_2 solution using another dropper.
- Add 3 drops of chlorine water to the content of the test tube and shake well.

Result:

- CS_2 forms a separate layer above the Lassaigne's extract.
- On adding chlorine water, the sodium bromide (NaBr) formed during the lassaigne's extract preparation reacts with chlorine water, resulting in oxidation of bromide to bromine (which dissolves in CS_2), imparting an orange colour in the CS_2 layer.

(bii) Carbon disulphide test for iodine

Materials Required:
Chlorine water, carbon disulphide (CS_2), dilute HCl, Lassaigne's extract, test tubes and droppers.

Procedure:

- Take 1 mL of Lassaigne's extract into a clean test tube using a dropper.
- Add 1 mL of CS_2 solution using another dropper.
- Add 3 drops of chlorine water to the content of the test tube and shake well.

Result:

- CS_2 forms a separate layer above the Lassaigne's extract.

- On adding chlorine water, the sodium iodide (NaI) formed during the lassaigne's extract preparation reacts with chlorine water, resulting in oxidation of bromide to bromine (which dissolves in CS_2), imparting a violet colour in the CS_2 layer.

CHAPTER

THREE

FUNCTIONAL GROUP TESTS AND ANALYSES.

1. Test for Alcohols

Alcohols are organic compound containing hydroxyl (-OH) functional group. Organic alcoholic compounds containing one –OH group are termed as monohydric, dihydric for two –OH groups and trihydric for three –OH groups. They are further classified as primary (CH_3CH_2OH), secondary ($CH_3CHOHCH_3$) and tertiary ($CH_3COHCCH_3CH_3$) alcohols based on the respective number of the R – groups attached to the carbon atom bearing the hydroxyl group.

(a) Sodium metal test

Materials Required:

Organic compound, anhydrous calcium sulphate ($CaSO_4$), sodium metal, test tube, glass funnel, filter papers, spatula and forceps.

Procedure:

- Take a small quantity of organic compound into a clean test tube using a spatula.
- Add small quantity of anhydrous $CaSO_4$ and shake the test tube well.
- Filter off the content using a glass-funnel fitted with filter paper into another clean test tube to obtain a filtrate.
- Using forceps take a small piece of sodium metal into a flat filter paper, fold and press to dry the sodium metal off.
- Put the dried sodium metal into the filtrate solution using forceps.

Result:

- There is production of effervescence of hydrogen gas.

$$2R\text{-}OH + 2Na \longrightarrow 2R\text{-}O^- Na^+ + H_2 (g)$$

Sodium alkaoxide

(b) Ester test

Materials Required:

Organic compound, glacial acetic acid, concentrated sulphuric acid (H_2SO_4), cold water, test tube, droppers, beaker and water bath (use beaker of water placed on a tripod stand under Bunsen burner flame as an alternative).

Procedure:

- Take small quantity of organic compound into a clean test tube.
- Add few drops of glacial acetic acid using a dropper.
- Add few drops of Conc. H_2SO_4 using another dropper.
- Warm the mixture in a water bath for few minutes.
- Pour the content of the mixture into a beaker containing 20 mL of cold water.
- Smell by wafting.

Result:

- Formation of a fruity smelly compound called ester.

$$ROH + R'COOH \xrightleftharpoons{\text{Conc. } H_2SO_4} R'COOR + H_2O$$

Ester

(c) **Ceric ammonium nitrate test**

Materials Required:

Organic compound, ceric ammonium nitrate, test tube, and droppers.

Procedure:

- Take small quantity of organic compound into a clean test tube using a dropper.
- Add few drops of ceric ammonium nitrate using another dropper.

Result:

- Formation of red coloured complex.

$$2ROH + (NH_4)_2Ce(NO_3)_6 \longrightarrow (ROH)_2Ce(NO_3)_4 + 2NH_4NO_3$$

Alkaoxycerium(IV) complex

(d) Acetyl chloride test

Materials required:

Organic compound, anhydrous calcium carbonate (CaSO₄), acetyl chloride, ammonium hydroxide solution (NH₄OH), test tubes, droppers, spatula, glass funnel, filter paper and glass rod.

Procedure:

- Take small quantity of organic compound into a clean test tube.
- Add few grams of anhydrous $CaSO_4$ using a spatula.
- Shake well to remove any water content and filter off in a glass funnel-fitted filter paper into a clean test tube.
- Add few drops of acetyl chloride and shake well.

- Deep a glass rod into NH$_4$OH solution and bring it close to the mouth of the test tube.

Result:

- Alcohol reacts with acetyl chloride to liberate hydrogen chloride gas.

$$ROH + CH_3COCl \longrightarrow CH_3COOR + HCl\ (g)$$

- The hydrogen chloride gas released reacts with NH$_4$OH to form white fumes of ammonium chloride (NH$_4$Cl).

$$HCl + NH_4OH \longrightarrow NH_4Cl + H_2O$$

(e)   Iodo form test

Materials required:

Organic compound, 1% iodine solution, dilute NaOH, test tube, droppers and water bath (use beaker of water placed on a tripod stand under Bunsen burner flame as an alternative).

Procedure:

- Take small quantity of organic compound into a clean test tube.
- Add few drops of 1% iodine solution using a dropper.
- Add dil. NaOH dropwise until the brown colour of iodine disappears.
- Warm the content mixture gently in a water bath.

Result:

- A yellow precipitate of iodoform is formed

$$CH_3CH(OH)CH_3 + I_2 + 2NaOH \longrightarrow CH_3(CO)CH_3 + 2NaI + 2H_2O$$

$$CH_3(OH)CH_3 + 3I_3 + 4NaOH \longrightarrow CHI_3 + CH_3COONa + 3NaI + 3H_2O$$
$$\text{Iodoform}$$

(f)   Lucas' test

Materials Required:

Primary alcohol, Secondary alcohol, tertiary alcohol, Lucas' reagent, droppers, test tubes, water bath (use beaker of water placed on a tripod stand under Bunsen burner flame as an alternative).

Procedure:

- Take small quantities of primary, secondary and tertiary alcohols into test tubes A, B and C respectively.
- Add few drops of Lucas' reagent in test tube C and B, using a dropper and shake well, and on test tube A after adding Lucas' reagent warm on a water bath.

Result:

- Tertiary alcohol reacts instantaneously with Lucas' reagent to form insoluble, cloudy alkyl chloride.

$$H_3C-\underset{\underset{CH_3}{|}}{\overset{\overset{CH_3}{|}}{C}}-OH \xrightarrow{ZnCl_2 \ + \ HCl} H_3C-\underset{\underset{CH_3}{|}}{\overset{\overset{CH_3}{|}}{C}}-Cl$$

- Secondary alcohol reacts in few minutes with Lucas' reagent to form insoluble, cloudy alkyl chloride.

$$H_3C-\underset{\underset{H}{|}}{\overset{\overset{OH}{|}}{C}}-CH_2CH_3 \xrightarrow{ZnCl_2 \ + \ HCl} H_3C-\underset{\underset{H}{|}}{\overset{\overset{Cl}{|}}{C}}-CH_2CH_3$$

- Primary alcohol reacts with Lucas' reagent on warming to form insoluble, cloudy alkyl chloride.

$$CH_3CH_2OH \xrightarrow{ZnCl_2 \ + \ HCl} CH_3CH_2Cl$$
$$\text{Ethyl Chloride}$$

2. Test for Phenols

These are compounds containing one or more hydroxyl groups attached to aromatic ring. Examples are cresol, catechol, resorcinol, etc. They are usually white crystalline solid.

(a) Litmus test

Materials required:
Organic compound, moist blue litmus paper, spatula and watch glass.

Procedure:

- Take a small quantity of organic compound and place in a moist blue litmus paper placed on a watch glass.

Result:

- Phenol is a weak acid and it turns the colour of the moist blue litmus paper red.

(b) Ferric chloride test

Materials required:
Organic compound, ferric chloride (neutral) solution, spatula, test tube and droppers.

Procedure:

- Take small quantity of neutral solution of ferric chloride into a clean test tube.
- Add small quantity of organic compound into the solution.

Result:

- Formation of a violet colour of a complex.

$$6C_6H_5OH + FeCl_3 \longrightarrow Fe[(OC_6H_5)_6]^{3-} + 3HCl\ 3H^+$$

(c) Libermann's test

Materials required:

Organic compound, sodium nitrite ($NaNO_2$), Conc. H_2SO_4, distilled water, NaOH solution, spatula, test tube, droppers, Bunsen and water bath.

Procedure:

- Take small quantity of $NaNO_2$ into a clean test tube with a spatula.
- Add small quantity of organic compound to the test tube with a dropper and heat gently for few minutes over Bunsen burner flame.
- Remove, allow to cool and using another dropper, add few drops of conc. H_2SO_4 and shake well.
- Add few milliliters of distilled water and shake.
- Add excess NaOH solution and shake well.

Result:

- There is a formation of deep blue or green colour product.

$$HO\text{-} \xrightarrow[HNO_3]{H_2SO_4} ON\text{-}\text{-}OH \rightleftharpoons HON\text{=}\text{=}O$$

p-nitrosophenol

- On addition of distilled water to the product it changes to red or brown colour of indophenol.

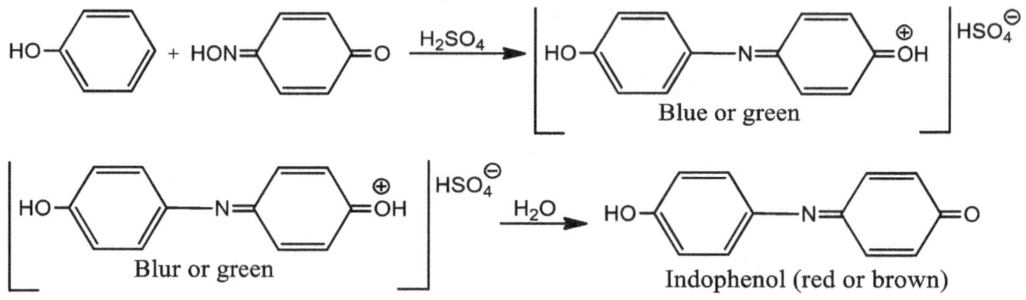

- On addition of NaOH solution, the red or brown colour changes back to blue or green colour product of indophenol anion.

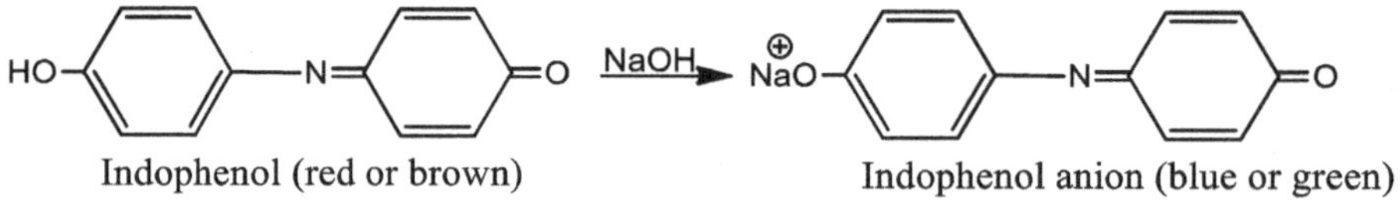

(d) Phthalein dye test

Materials required:

Organic compound, phthalic anhydride, conc. H_2SO_4, dilute NaOH, spatula, beaker, test tube, dropper and water bath (use beaker of water placed on a tripod stand under Bunsen burner flame as an alternative).

Procedure:

- Take small quantity of organic compound into a clean test tube.
- Add some quantity of phthalic anhydride using a spatula.
- Using a dropper add few drops of conc. H_2SO_4 to the content of the test tube.
- Heat the content mixture for few minutes in a water bath and cool in a beaker of cold water.

- Pour the reaction mixture into a beaker containing about 20 mL of dilute NaOH solution.

Result:

- On heating, Phenolphthalein (colourless) is formed.

$$2 \text{ phenol} + \text{anhydride} \xrightarrow{H_2SO_4} \text{phenolphthalein (colourless)}$$

phenolphthalein (colourless)

- On reaction with dilute NaOH, phenolphthalein solution turns pink.

$$\text{phenolphthalein (colourless)} \xrightarrow{NaOH} \text{pink colour}$$

phenolphthalein (colourless)   pink colour

Precaution:

- Phenols should be handled with care because it is corrosive and toxic.
- Concentrated H_2SO_4 is hazardous and causes skin burn, therefore it should be handled with care.

3. **Test for Aldehydes**

   Aldehydes are organic compounds in which the carbonyl carbon is attached to hydrogen and R-group (R = alkyl or aryl group). Aldehydes are presence in almond, cinnamon, etc.

(a) 2,4-Dinitrophenyl hydrazine test

Materials required:

Organic compound, rectified spirit, 2,4-dinitrophenyl hydrazine solution, test tube and droppers.

Procedure:

- Take a small quantity of organic compound into a clean test tube.
- Add few drops of rectified spirit using a dropper and shake well.
- Add few drops of 2,4-dinitrophenyl hydrazine solution and shake well.

Result:

- Formation of a yellow or orange precipitate of 2,4-dinitrophenyl hydrazone.

$$H_3C-\overset{\overset{H}{|}}{C}=O \ + \ H_2N-N(H)-C_6H_3(NO_2)(O_2N) \longrightarrow H_3C-\overset{\overset{H}{|}}{C}=N-N(H)-C_6H_3(NO_2)(O_2N)$$

Acetaldehyde-2,4-dinitrophenyl hydrazone

(b) Sodium bisulphite

Materials required:

Organic compound, sodium bisulphite solution (saturated), droppers, cork and boiling tube.

Procedure:

- Take a small amount of saturated solution of sodium bisulphite into a clean boiling tube.
- Add small quantity of organic compound into the content mixture using a dropper.
- Cork the boiling tube and shake well.

Result:

- Formation of white crystalline bisulphite addition product.

$$R-\overset{\overset{\displaystyle H}{|}}{C}=O \ + \ NaHSO_3 \longrightarrow R-\overset{\overset{\displaystyle H}{|}}{\underset{\underset{\displaystyle SO_3Na}{|}}{C}}-OH$$

Bisulphite addition product

(c) Schiff's test.

Materials required:
Organic compound, Schiff's reagent, test tube and droppers.

Procedure:

- Take a small quantity of organic compound into a clean test tube.
- Add few drops of Schiff reagent using a dropper.

Result:

- Pink-magenta colour develops, showing presence of aldehyde.

Schiff's
Reagent

(d) Tollen's test.

Materials required:

Organic compound, silver nitrate ($AgNO_3$) solution, NH_4OH solution, dilute NaOH, test tube, dropper and water bath (use beaker of water placed on a tripod stand under Bunsen burner flame as an alternative).

Procedure:

- Take a small quantity of $AgNO_3$ solution into a clean test tube using a dropper.

- Add few drops of dilute NaOH using another dropper.
- Add excess of dilute NH_4OH to the precipitate formed using a dropper and shake well.
- Add few drops of organic compound to the test tube using another dropper.
- Heat the content mixture in a boiling water bath.

Result:

- On addition of dilute NaOH to $AgNO_3$ solution, black precipitate of silver Oxide is formed.
- On heating the content mixture, silver metal is formed on the inner surface of the test tube.

$$RCHO + \boxed{2Ag(NH_3)_2^+ + 3OH^-} \longrightarrow RCOO^- + 2Ag\,(s) + 4NH_3 + 2H_2O$$

$$\underset{\text{Tollen's reagent}}{} \qquad \qquad \underset{\text{silver}}{}$$

(e) Fehling's test

Materials required:

Organic compound, Fehling's solution A, Fehling's solution B, test tube, droppers and water bath (use beaker of water placed on a tripod stand under Bunsen burner flame as an alternative).

Procedure:

- Take small quantity of organic compound into a clean test tube.
- Add small quantity of equal amount of Fehling's solutions A and B using different droppers.
- Heat the content mixture in a boiling water bath.

Result:

- Red precipitate of copper(I)oxide is formed which is as a result of reduction of blue copper(II)oxide present in the test reagent's mixture.

$$RCHO + [2Cu^+ + 3OH^-] \longrightarrow RCOO^- + Cu_2O\ (s) + 2H_2O$$

Fehling's solution, copper(I)oxide (red)

4. Test for Ketones

Ketones are organic compounds containing carbonyl carbon that is attached to two R-groups, that is, R and R' groups. The R-groups can be alkyl or aryl groups. The presence of ketones is observed in the flavours of berries, mushrooms, etc.

(a) 2,4-Dinitrophenyl hydrazine test

Materials required:

Organic compound, rectified spirit, 2,4-dinitrophenyl hydrazine solution, test tube and droppers.

Procedure:

- Take a small quantity of organic compound into a clean test tube.
- Add few drops of rectified spirit using a dropper and shake well.
- Add small quantity of 2,4-dinitrophenyl hydrazine solution into the content of the test tube using another dropper and shake well.

Result:

- Formation of orange or yellow precipitate of 2,4-dinitrophenyl hydrazone.

2,4-dinitrophenyl hydrazone   Acetone-2,4-dinitrophenyl hydrazone

(b) **Sodium bisulphite**

Materials required:

Organic compound, sodium bisulphite solution (saturated), droppers, cork and test tube.

Procedure:

- Take a small amount of saturated solution of sodium bisulphite into a clean test tube.
- Add few quantities of the organic compound into the content mixture using a dropper.
- Cork the boiling tube and shake well, leave for few minutes.

Result:

- Formation of white crystalline bisulphite addition product.

$$R\text{--}\underset{\underset{R}{|}}{C}\text{=}O \;+\; NaHSO_3 \longrightarrow R\text{--}\underset{\underset{SO_3\overset{+}{Na}}{|}}{\overset{\overset{R}{|}}{C}}\text{--}\overset{-}{O}$$

Bisulphite addition product

(c) **Meta-dinitrobenzene test**

Material required:

Organic compound, meta-dinitrobenzene, dilute NaOH, test tube, dropper and spatula.

Procedure:

- Take small quantity of organic compound into a clean test tube.
- Using a spatula add small quantity of meta-dinitrobenzene.
- Add few drops dilute NaOH using a dropper and shake the test tube well.

Result:

- A violet colouration that slowly fades away is observed.

(d) Sodium nitroprusside test

Materials required:

Organic compound, sodium nitroprusside crystal, dilute NaOH, distilled water, test tube and dropper and spatula.

Procedure:

- Take a small quantity of the crystals of sodium nitroprusside into a clean test tube using a spatula.
- Add small amount of distilled water and shake well until the crystal is dissolved.
- Add small quantity of organic compound using a dropper.
- Add few drops dilute NaOH using another dropper and shake.

Result:

- Formation of red coloured complex.

$$CH_3COCH_3 + OH^- \longrightarrow CH_3COCH_2^- + H_2O$$
$$\text{Ketone} \qquad\qquad\qquad \text{ketone anion}$$

$$[Fe(CN)_3NO]^{2-} + CH_3COCH_2^- \longrightarrow [Fe(CN)_3NO.CH_3COCH_2]^{2-}$$
$$\text{Nitroprusside ion} \qquad\qquad\qquad \text{(red coloured complex)}$$

5. Test for Carboxylic Acids

Carboxylic acids are organic compounds containing carboxyl functional group (-COOH). Aliphatic and aromatic carboxylic acids are the two types of carboxylic acid. Examples are acetic acid, formic acid and benzoic acid, etc. Carboxylic acids are also found in fruits such as orange, apple, lemon, etc.

(a) Litmus test.

Materials required:

Organic compound, moist blue litmus paper, dropper and watch glass.

Procedure:

- Take few drops of organic compound using a dropper into a moist blue litmus paper placed on a watch glass.

Result:

- The blue litmus paper turns red, showing it is acidic in nature.

(b) Sodium bicarbonate test

Material required:

Organic compound, sodium bicarbonate, spatula, test tube and dropper.

Procedure:

- Take small quantity of organic compound into a clean test tube using a dropper.
- Add small quantity of sodium bicarbonate to the compound using a spatula.

Result:

- Brisk effervescence of carbon(IV)Oxide gas is produce.

$$RCOOH + NaHCO_3 \longrightarrow RCOONa + CO_2\,(g) + H_2O$$

(c) Ester test

Materials required:

Organic compound, ethyl alcohol, conc. H_2SO_4, distilled water, test tube, droppers and water bath (use beaker of water placed on a tripod stand under Bunsen burner flame as an alternative).

Procedure:

- Take small quantity of organic compound into a clean test tube.
- Add few drops of ethyl alcohol into the test tube using a dropper.

- Add few drops of conc. H_2SO_4 into the content mixture.
- Heat the reaction mixture in water bath for few minutes.
- Pour the content mixture into a beaker containing 20 mL of distilled water and smell by wafting.

Result:

- Production of a fruity smelly compound called ester.

$$R'COOH + R'OH \xrightarrow{\text{Conc. } H_2SO_4} \underset{\text{Ester}}{R'COOR} + H_2O$$

6. Test for Amines

Amines are derived from ammonia by replacing the hydrogen atom(s) with R- group(s). The R group(s) can alkyl or aryl groups. Primary, secondary and tertiary amines are the three types of amine. When the one hydrogen of ammonia is replaced by one R-group primary amine is formed; when the two hydrogens are replaced with two R-groups, secondary amine is formed; when all the three hydrogens are replaced with R-groups, tertiary amine is formed.

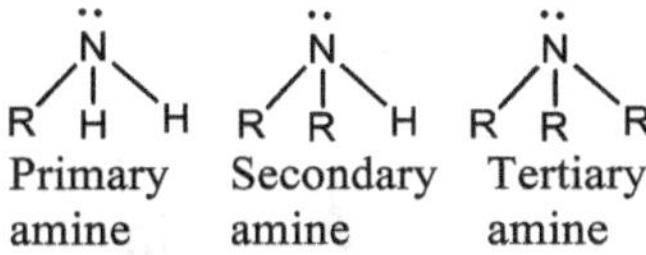

Examples are Nicotine (present in tobacco), Caffeine (present in coffee plant), etc.

(a) Solubility test

Materials required:

Organic compound, dilute HCl, droppers and test tubes.

Procedure:

- Take small quantity of organic compound into a clean test tube.
- Add small quantity of dilute HCl using a dropper and the test tube well.

Result:

- Dissolution to form a solution, showing that amines are basic in nature.

(b) Litmus test

Materials required:
Organic compound, moist red litmus paper, dropper and watch glass.

Procedure:

- Take few drops of organic compound using a dropper into a moist red litmus paper placed on a watch glass.

Result:

- The red litmus paper turns blue, showing that amine is basic in nature.

(c) Azo-dye test.

Material required:
Organic compound, sodium nitrite ($NaNO_2$), dilute HCl, distilled water, dilute NaOH, ice cold water, test tubes and droppers.

Procedure:

- Take small quantity of organic compound, $NaNO_2$, 2-naphthol into clean test tubes A, B and C respectively.
- Add small quantity of dilute HCl to test tube A using a dropper and shake well.

- Add small quantity of distilled water to test tube B from distilled water bottle and shake well until a solution is formed.

- Add small quantity of dilute NaOH solution to test tube C using a dropper and shake well until a solution is formed.

- Cool the test tube A, B and C in a beaker containing ice cold water for 5 minutes.

- Add the solution in test tube B into the solution in test tube A to form a reactive mixture.

- Add the reactive mixture to solution in test tube C.

Result:

- Formation of diazonium salt on mixing solution B with solution A after ice cold water cooling.

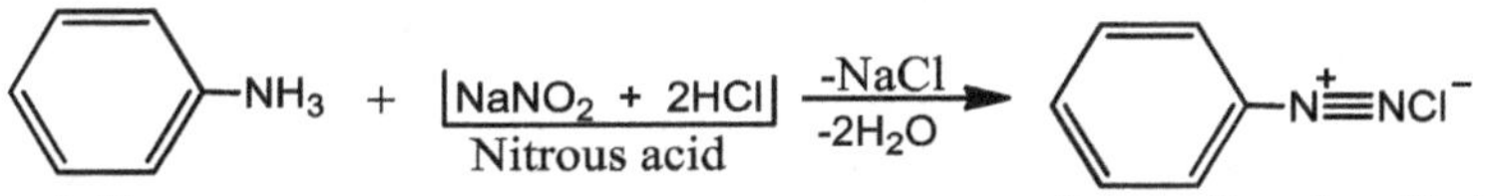

$$C_6H_5-NH_3 + [NaNO_2 + 2HCl] \xrightarrow[-2H_2O]{-NaCl} C_6H_5-N\equiv NCl^-$$

Aniline   Nitrous acid   Benzenediazonium chloride

- Production of a scarlet red coloured compound called azo-dye, on reaction of diazonium salt mixture to the cooled solution in test tube C.

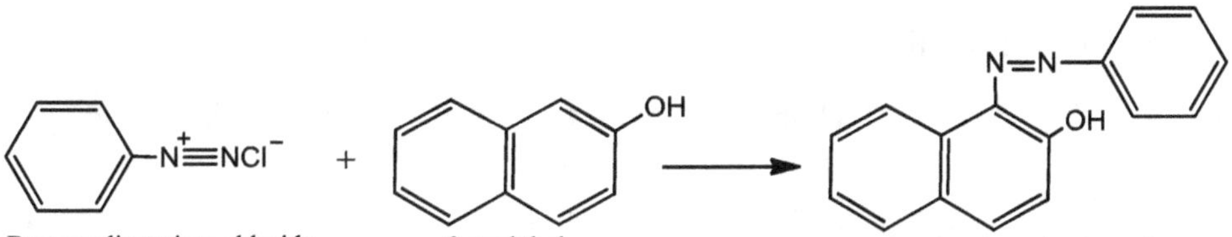

Benzenediazonium chloride   2-naphthol   Azo dye (scarlet red coloured)

(d) Nitrous acid test

Materials required:

Primary amine, secondary amine, tertiary amine, $NaNO_2$, distilled water, conc. HCl, ice cold water, test tubes, dropper and spatula.

Procedure:

- Take small quantities of primary, secondary and tertiary amines into clean test tubes A, B and C respectively.

- Add small quantity of $NaNO_2$ into another clean test tube using a spatula, and add small quantity of distilled water to dissolve it and shake to get $NaNO_2$ solution.

- Cool the $NaNO_2$ solution in a beaker containing ice cold water for 5 minutes.

- Add small quantity of conc. HCl each to test tubes A, B and C respectively using a dropper and shake the test tubes well.

- Cool the test tubes, A, B and C solutions in the ice cold water for 5 minutes.

- Add small quantity of the cooled $NaNO_2$ solution to each of the cooled solutions in test tubes A, B and C respectively.

Result:

- A bubble of Nitrogen gas is produced in test tube A.

$$RNH_2 \quad + \quad HONO \quad \longrightarrow \quad ROH + H_2O + N_2\ (g)$$

1^0 Amine   Nitrous   Alcohol   Nitrogen

(Aniline)   acid   gas

- An oily yellow nitrosamine is produced in test tube B.

$$R_2NH \quad + \quad HONO \quad \longrightarrow \quad R_2N-NO \quad + \quad H_2O$$

2^0 Amine   Nitrous   Nitrosamine

                 acid

- Formation of a soluble salt is observed in test tube C.

$$R_3N \quad + \quad HONO \quad \longrightarrow \quad R_3\overset{+}{N}H-O\overset{-}{N}O$$

3^0 Amine   Nitrous   Trialkyl ammonium

                 acid   nitrite (soluble salt)

(e) Hinsburg's test

Materials required:

Primary amine, secondary amine, tertiary amine, 25% NaOH solution, distilled water, benzene sulphonyl chloride, conc. HCl, beaker (containing 20 mL of water), test tube and droppers.

Procedure:

- Take small quantities of primary, secondary and tertiary amines into clean test tubes A, B and C respectively.

- Add small quantities of 25% NaOH solution each to test tubes A, B and C respectively using a dropper and shake well.
- Add small quantities of distilled water to each of the test tubes A, B and C respectively.

- Add small quantities of benzene sulphonyl chloride solution each to the test tubes A, B and C respectively using another dropper.

- Shake the mixtures in each of the test tubes A, B and C and cool in a beaker containing water for few minutes.

- Add small quantities of conc. HCl into each of test tubes A, B and C respectively using a dropper.

Result:

- Soluble sulphonamide is formed in test tube A which on reaction with conc. HCl forms insoluble sulphonamide.

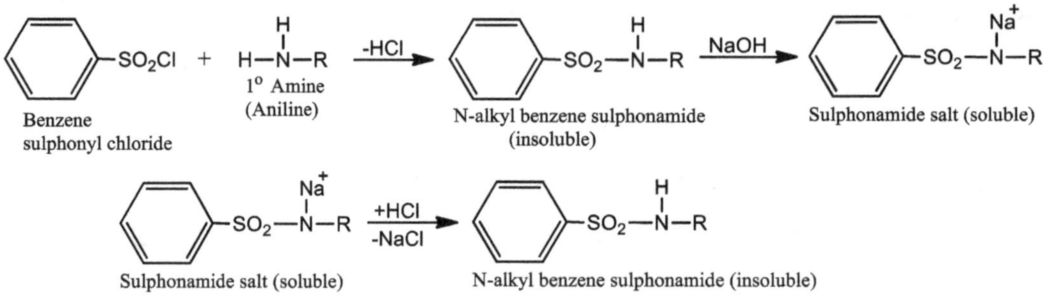

- Insoluble sulphonamide is formed in test tube B which will not react with conc. HCl.

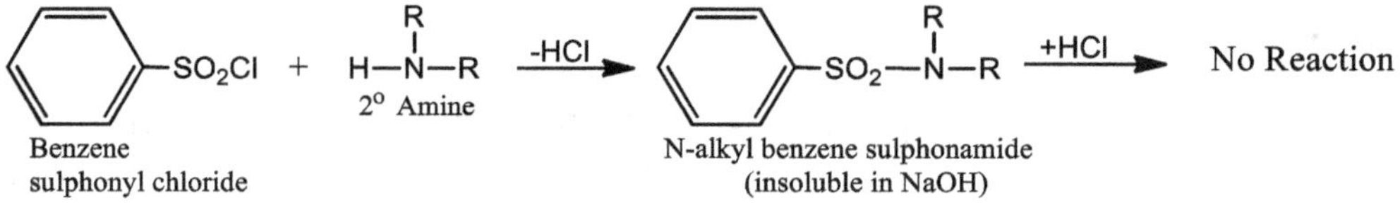

- No reaction with benzene sulphonyl chloride is observed in test tube C. it remains insoluble in NaOH solution but will react with conc. HCl to form soluble ammonium salt.

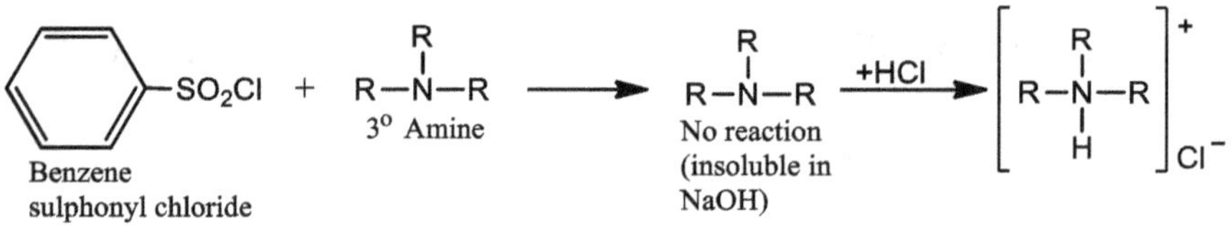

Precaution:

- Handle the organic compounds with proper care.

- Handle the reagents with care and avoid contamination.

- Handle the acids and bases with proper care because they are corrosive and can cause skin burns.

7. Test for Cyanides

Organic compounds containing cyanide has CN bond attached in their structure. Cyanide could be in solid, liquid or gaseous forms. They are poisonous substances and should be handle carefully.

Test for cyanide can be carried out as follows:

Material required:

Sodium hydroxide (NaOH) solution (0.5 M). Ferrous sulphate ($FeSO_4$) solution (1 M), dilute HCl (0.5 M), unknown compound, test tube and droppers

Procedure:

- Add small quantity of the unknown compound into a test tube.
- Add one drop of dil. NaOH into the test tube containing the compound using a dropper.
- Add 15 drops of Ferrous sulphate ($FeSO_4$) solution into the mixture using another dropper and shake well.
- Add to the content mixture 5 drops using a different dropper and observe.

Result:

- Formation of dark cherry red colour.

CHAPTER

FOUR

BIOSYNTHESIS OF ORGANIC COMPOUND

Organic compounds can also be produced from natural means or biological species, such as microorganisms, plants and animals. Several hundreds of the known phenazines have been isolated from natural sources, mostly *Pseudomonas* strains, other gram-negative and gram-positive bacteria.

(a) Production of Phenazine-1-carboxylic acid (PCA) from microorganism

Materials required:

Pseudomonas aeruginosa, Luria-Bartani (LB) agar, King's A broth (bactopeptone (15.0 g), sodium chloride (13.0 g), glycerol (9.0 mL) and potassium sulphate (1.0 g) in 1000 mL distilled water), Orbital shaker, Erlenmeyers flask, centrifuge, HCl, dichloromethane, ethyl acetate, acetonitrile, Amberlite XAD-16 resin, silica gel, column and distilled water.

Procedure:

- Streak a pure *Pseudomonas aeruginosa* culture on Luria-Bertani (LB) agar plate and incubate at room temperature for 24 hrs.
- Transfer the colony of *P. aeruginosa* formed on a LB agar plate into Erlenmeyer flask containing 100 mL King's A broth and incubate at 29 - 30°C with an orbital shaker (200 rpm) for 24 hrs.

- Use an Amberlite XAD-16 resin column for the PCA isolation by eluting this column with 70 % (v/v) acetonitrile in distilled water.

- For purification purposes, adjust the pH of the crude phenazine solution with HCl to 2.5 and remove the residues by centrifugation at 3,500 rpm for 15 minutes.

- Extract the PCA from the solution with dichloromethane.

- For further purification on a silica gel column, equilibrate the extracted PCA with 90% (v/v) dichloromethane in ethyl acetate.

Result:

- A yellow crystal of PCA concentration is formed.

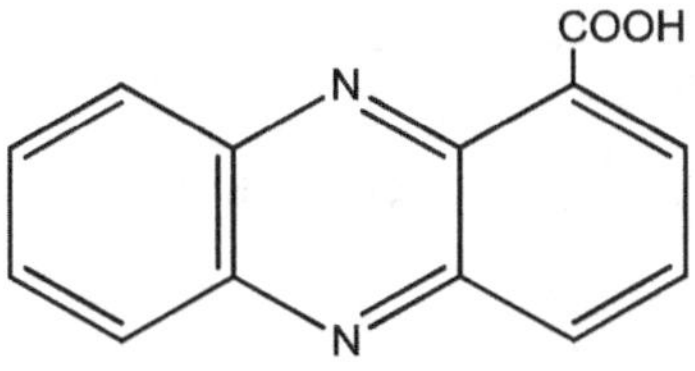

Phenazine-1-carboxylic acid
(PCA)

CHAPTER

FIVE

PREPARATION OF SOLUTION AND CALCULATION EQUATIONS

In the course of experiment, chemical solutions of compounds are calculated and prepared to the concentrations required before they are used for tests, reactions and analyses.

(a)   Preparation method of solution in distilled water using solid chemical compound.

Procedure is as follows:

- Calculate the amount of solute required in 1 Liter of the solvent (distilled water or deionized) at the required molarity.
- Weigh out with watch glass using a weighing balance.
- Transfer to a calibrated beaker containing half (0.5 L) volume of the solvent.
- Wash of the remaining particles of the solute attached to the watch glass with few volume of the solvent.
- Stir with a stirrer (glass rod).
- Apply small heat to assist fast dissolution in case of delay dissolution.
- Transfer the content of the beaker to a volumetric flask or a measuring cylinder using glass funnel.
- Rinse the beaker, glass rod and glass funnel into the flask with few amount of the solvent.
- Make up the solution in the flask with the solvent up to 1liter mark (at the meniscus).
- Cork flask with a stopper and mix thoroughly.
- Transfer to the labelled reagent bottle.

(b)   Determination of concentration by percentage of substances

This can be expressed as percentage weight/volume (% w/v) or percentage volume/volume (% v/v) or percentage weight/weight (% w/w).

For solid solute:

Percentage weight per volume (% w/v) is used and expressed as the mass of solute (solid) in grams dissolved in 100 mL of solution.

$$\% \, w/v = \frac{mass \; of \; solute \; (g)}{volume \; of \; solution \; (mL)} \; x \; 100\%$$

For liquid solute:

Percentage volume per volume (% v/v) is used and expressed as the volume of solute (liquid) in milliliter per 100 mL of solution.

$$\% \, v/v = \frac{volume \; of \; solute}{volume \; of \; the \; solution} \; x \; 100\%$$

Weight percent (% w/w):

This is the mass of solute (in gram) in grams (100 g) of solution. The solution can be prepared independent of the temperature considerations, usually in a commercial preparation.

$$\% \, w/w = \frac{mass \; of \; solute}{mass \; of \; solution} \; x \; 100\%$$

(c)   Determination of percentage yield

In the course of an experiment, the yield or product in percentage is calculated using the practical yield against the theoretical yield. It is assumed that the materials used in the practical react in such a way that the quantity expected as product of the reaction may decrease from the stoichiometric quantity evaluated.

$$\% \; yield = \frac{practical \; (g)}{theoretical \; yield \; (g)} \; x \; 100\%$$

(d)   Preparation of solution by dilution

In dilution, more solvent (diluent) is added to a given solution. Solutions can be prepared by dilution. A solution of higher concentration is diluted to produce solutions with lower concentrations. A stock solution is a concentrated solution that is diluted to a lower concentration, also called working solution.

This dilution does not alter the number of moles of solute present as they are the same before and after the dilution but alters only the final volume.

$$Concentration \; of \; solution \; (C) \; in \; mol/L = \frac{number \; of \; moles \; of \; the \; solute \; (n)}{volume \; of \; the \; solution \; (v) \; in \; Liter}$$

$$n = Cv$$

Therefore, for the two dilutions: initial dilution (1) which is the concentrated solution and final dilution (2) which is the diluted solution:

$$C1V1 = C2V2$$

$$V1 = \frac{C2V2}{C1}$$

(e) Preparation of a molar solution

Molarity is the number of moles of solute per liter (1000 mL) of solution. In preparation of saturated solution of a liquid solute, the number of volume of solvent (usually distilled or deionized water) to be used in such preparation is calculated. The known values required are the molarity of the substance to be prepared (M), specific gravity or density, percentage assay and molecular mass. The figures are usually written on the body of the container of the commercial chemical product. Note that, averagely concentrated Hydrochloric acid (HCl) has a molecular mass of 36.5 g/mol, specific gravity of 1.18, and percentage assay, 36%.

Mathematically,

$$Volume\ in\ 1\ L\ for\ 1\ M = \frac{concentration\ (M)\ x\ molecular\ mass\ (g/mol)}{percentage\ assay\ x\ specific\ gravity}\ x\ 100\%$$

Most liquid concentrated chemical solutions have their molecular mass, specific gravity and percentage assay data written on the bottle or container labels.

INDEX